HOUSE OF ABUNDANCE
PUBLICATIONS

Simulation Theory for Beginners: Evaluating the Simulation Hypothesis and Its Virtual Reality Matrix

"If real is what you can feel, smell, taste and see, then 'real' is simply electrical signals interpreted by your brain."

— MORPHEUS, THE MATRIX

Contents

1

Introduction

In a world riddled with technological marvels, a compelling question emerges as our screens flood images of virtual realities and artificial intelligence: Are we living in a simulation? This thought projects itself into our consciousness, inviting us to entertain the idea that our reality might not be as 'real' as we once believed. This is the realm of the Simulation Theory, a hypothesis suggesting that our reality might be a computer-generated construct, not unlike the virtual worlds in video games or VR simulations.

Defining Simulation Theory: What is it?

The Simulation Theory, or the Simulation Hypothesis, posits that all reality, including Earth and the universe, is an artificial construct. In this model, we, as conscious beings, might be part of this virtual simulation without any genuine awareness of our actual reality. Just as characters in a video game are unaware of the player controlling their actions, we might be the characters in a sophisticated, hyper-realistic game played by entities in a 'base reality.'

Drawing parallels with the virtual realities we create, think of your favorite video game. Within this game, characters operate, landscapes evolve, and events happen. Now, imagine a game so advanced that the characters within it believe they are alive and have emotions, dreams, and aspirations. The Simulation Theory suggests that our universe might be such a game, an intricate digital creation where we're the characters, and our perception of reality is nothing more than a series of programmed events.

Historical Origins: Where did it come from?

While technological advancements and philosophical arguments bolster the modern version of the Simulation Theory, its roots can be traced back to ancient philosophical thought. Philosophers have always grappled with the nature of reality and our perception of it.

One of the earliest and most famous explorations of a 'simulated' reality is Plato's Allegory of the Cave. Here, people are chained inside a cave and can only see the shadows cast on the wall by objects behind them. For them, these shadows are the only reality they know. This allegory questions the nature of reality and suggests that our perceptions might be mere shadows of a higher, more genuine reality.

Fast forward to the Enlightenment era, and we find René Descartes wrestling with a similar theme. He posited that there's no surefire way to confirm that the external world is not a dream or illusion. With the phrase "Cogito, ergo sum" (I think, therefore I am), Descartes acknowledged that the only undeniable truth is one's consciousness and existence.

The contemporary formulation of the Simulation Theory owes much to Dr. Nick Bostrom's 2003 paper "Are We Living in a Computer Simulation?" Here, he presented a statistical argument suggesting that if it's possible to create realistic simulations of the universe, then it's statistically probable that we're in one.

Importance: Why Should We Care?

Brushing off the Simulation Theory as a mere intellectual exercise or a fodder for sci-fi enthusiasts would be premature. If true, the implications are immense, affecting our understanding of existence, consciousness, and the universe.

Firstly, the nature of existence takes center stage. If we're in a simulation, what does it mean to 'exist'? It challenges our fundamental notions of reality, pushing us to reconsider our place in the universe. Are we mere lines of code in a cosmic program? Or are we conscious entities with genuine emotions and experiences, even if our environment is artificial?

Next, there's the matter of purpose. If our reality is a construct, it begs the question: Why was it created? Are we part of a cosmic experiment, entertainment for higher beings, or perhaps a way for these entities to explore their past?

Additionally, the Simulation Theory has practical implications.

If our universe operates based on specific algorithms or codes, understanding these could lead to groundbreaking technological and scientific advancements. It might enable us to 'hack' our reality, pushing the boundaries of what's possible.

To traverse the landscape of the Simulation Theory is to embark on a journey that merges philosophy, technology, and spirituality. It compels us to question, wonder, and probe deeper into the fabric of our existence. Whether you're a skeptic or a believer, exploring this hypothesis promises to be a mind-bending adventure, and this chapter was just the beginning.

2

Philosophical Foundations

T hroughout history, humans have been drawn to ponder the nature of reality, existence, and consciousness. Philosophers have dealt with these questions, attempting to understand what is real and how we can be sure of our experiences. The Simulation Theory, with its meaningful implications, doesn't emerge from a vacuum. Its roots are anchored deep within philosophical traditions that have challenged our perception of reality. Three foundational ideas provide essential groundwork: Plato's Allegory of the Cave, Descartes' musings on truth, and the brain in a Vat thought experiment.

Plato's Allegory of the Cave: An Early Hint

The ancient Greek philosopher Plato provided one of the earliest and most vivid explorations of perceived reality in his work "The Republic." The Allegory of the Cave is a captivating narrative that challenges us to question what we believe to be accurate.

Imagine a group of people who have lived in a dark cave. Their movement is so restricted that they only gaze upon the cave wall before them. Behind them burns a fire, and a raised walkway is between the prisoners and the fire. Objects are paraded on this walkway, casting shadows upon the cave wall due to the fire's light. For the prisoners, these shadows are the only reality they've ever known. They've never fully seen the outside world, the objects, or even each other. These shadows, mere representations, are mistaken for reality.

One day, a prisoner is released. He is dragged out of the cave and into the sunlight. Initially, he is blinded and disoriented. But as his eyes adjust, he perceives a world of vibrant color, depth, and dimensionality, far more 'real' than the shadowy illusions of the cave. He returns to the shelter to liberate his fellow prisoners. Still, he struggles to convince them of a world beyond the shadows.

Plato's allegory is a deep meditation on the nature of perception, knowledge, and truth. The cave represents humans' limited reality, constrained by our senses and experiences. The outside world symbolizes a higher, more valid form of existence accessible through intellectual and philosophical pursuits.

Descartes and the Nature of Reality: Can We Trust Our Senses?

The Enlightenment philosopher René Descartes took the baton from Plato and ran with it, further probing the trustworthiness of our senses and the nature of reality. In his seminal work, "Meditations on First Philosophy," Descartes introduced the method of radical doubt. He systematically doubted everything he believed to be accurate, hoping to find something definite.

Descartes posited that our senses, at times, deceive us. We've all mistaken a distant stick for a snake or heard phantom sounds in a silent room. Suppose our senses can deceive us in such instances. How can we trust them to provide an accurate representation of reality?

To demonstrate the fragility of our sensory experiences, Descartes employed the dream argument. In a dream, everything feels as natural as in waking life. The experiences, emotions, and sensory inputs are indistinguishable from reality. We cannot reliably differentiate between waking life and dreams based on sensory information alone. How can we be sure that our current experiences aren't just a part of an elaborate dream Descartes' introspective journey led him to his famous conclusion: "Cogito, ergo sum" or "I think, therefore I am." While he doubted everything else, the very act of doubt proved his existence as a thinking entity.

The Brain in a Vat: A Thought Experiment

Emerging from the philosophical lineage that includes Plato's cave and Descartes' dream, the "Brain in a Vat" is a modern thought experiment that delves into similar themes but with a technological twist.

Imagine a scenario where a mad scientist removes your brain and places it in a vat filled with life-sustaining nutrients. Electrodes are attached to the brain, and a supercomputer sends electrical impulses to simulate experiences. These simulated experiences are indistinguishable from the ones you'd have if your brain were still in your body. In this scenario, how would you know if you're a brain in a vat or a person living a 'real' life? The central premise is that you wouldn't.

The "Brain in a Vat" thought experiment underlines the indeterminacy of our sensory experiences. Just as the shadows were the prisoners' reality in Plato's cave and dreams were indistinguishable from waking life for Descartes, the simulated brain experiences in the vat are its reality. It reinforces that our perceived reality is constructed from our sensory experiences, regardless of origin.

Navigating the corridors of philosophical inquiry, it becomes evident that the Simulation Theory is the latest in a series of attempts to understand reality and our place within it. From the chained prisoners in Plato's cave to Descartes' doubting self and the disembodied brain manipulated by technology, each narrative challenges us to question our assumptions, seek more profound understanding, and remain open to possibilities beyond our immediate perceptions. These foundations don't just provide historical context; they support our quest to fathom the enigmatic nature of existence.

3

Technology and Simulations

The Growth of Computer Technology: From 0s and 1s to Virtual Worlds

The dawn of the digital age ushered in an era of unprecedented transformation. The inception of computer technology, in its most rudimentary form, revolved around binary language—0s and 1s. This simple yet profound system of representation, originating from the concept of electrical switches being either off or on, laid the groundwork for a revolution.

Although a mammoth machine, the first programmable computer, the ENIAC provided glimpses of the potential inherent in automated computation. As time progressed, the hunger for miniaturization and enhancement took hold. Transistors, tiny semiconductor devices, replaced vacuum tubes, making computers more compact and efficient. The 1960s and 70s witnessed the birth of the microprocessor, a complete central processing unit on a single chip, transforming the landscape of technology and setting the stage for personal computers.

With the rise of the personal computer in the 1980s, computing became increasingly accessible. From large institutions to home offices, computers began changing the fabric of society. And as they evolved, so did their capacities. Simple arithmetic and data storage tasks expanded into intricate graphical user interfaces, multimedia capabilities, and connections with other computers.

The inception of the Internet was another watershed moment. Binary code was no longer just about calculations—it was now translating human communication, bridging geographical divides, and converting real-world entities into digital avatars.

Video Games and VR: Simulating Entire Universes

Suppose one were to seek a testament to the profound leaps in simulation capabilities. In that case, one needs to look at video games. The earliest video games were simple pixelated forms, yet they represented a novel way of interaction. Over the decades, games transitioned from 2D monochromatic blocks to 3D photorealistic worlds. The rudimentary Pong morphed into expansive realms like "The Witcher" series, where every leaf, every droplet of water, and every character expression bore an uncanny semblance to reality.

Virtual Reality (VR) pushed these boundaries even further. With VR, immersion was not just about visual representation but about experience. When one dons a VR headset, the line between the physical self and the digital avatar blurs. This technology offered more than just panoramic views; it provided the sensation of "presence." The wind on a virtual mountaintop, the sense of rain in a simulated forest, and the pressure of a handshake with a digital entity became experientially real.

Augmented Reality (AR), a cousin of VR, further bridged the gap between the tangible and the virtual. Instead of replacing reality, AR supplements it. Through AR glasses or smartphone screens, digital information overlays our physical surroundings, intertwining the real and the virtual in real time.

The Computational Power Needed for a Universe-Scale Simulation

The sheer magnitude of computational requirements can be astonishing when pondering a universe-scale simulation. Let's break this down: The human brain, with its approximately 86 billion neurons, is said to have an almost unfathomable computational capacity. Now, imagine simulating not just a single brain in its entirety but every brain, every organism, every atom, and every quantum event on Earth. Extend that to our solar system, galaxy, and beyond.

On the quantum level, the challenge deepens. Quantum mechanics, with phenomena like superposition and entanglement, operates counter to our intuitive sense of reality. To simulate the universe, one must account for these quantum quirks accurately.

Now, let's contemplate data storage. The Large Hadron Collider, one of the most complex machines humans have ever built, produces around 30 petabytes of data annually. This is just data from subatomic particle collisions. The storage requirement for an entire universe's worth of continuous data would be orders of magnitude greater.

But technology often finds ways to evolve and surprise. Quantum computing, a field in its infancy, promises to transcend classical computing limitations. Quantum computers have the potential to process vast amounts of information simultaneously using qubits instead of traditional bits, offering hope for solving previously insurmountable problems.

Moreover, if we look at the nature of simulations, not all data needs to be processed concurrently. Concepts like "rendering on demand," where the simulation only renders specific parts when they are observed (much like a video game only rendering what's in front of the player), could reduce the computational load.

Influence of Simulations on Society

While the technical aspects of simulations are intriguing, their influence on society is profound. Economies, for instance, are now intertwined with virtual currencies and digital assets. Real estate in virtual worlds sells for real money, and digital art pieces, represented as non-fungible tokens, auction for millions.

Simulations also redefine learning. Flight simulators train pilots without leaving the ground, and virtual labs allow students to conduct experiments in environments free from physical limitations.

Ethical implications arise, too. As simulations become indistinguishable from reality, questions about consent, privacy, and the very nature of existence gain prominence. Philosophers, ethicists, and technologists grapple with these issues, shaping the dialogue for the next generation.

In our exploration of simulations, we glimpse the milestones of technological evolution and the essence of human curiosity. As our simulations grow in scale and complexity, so does our understanding of the universe and our place within it. The dance between technology and exploration, between reality and the simulated, continues, promising a future prosperous with discovery and introspection.

4

The Proponents and Their Arguments

Nick Bostrom's Simulation Argument: The Trilemma

Nick Bostrom, a distinguished philosopher with a penchant for existential risks, has given the simulation hypothesis intellectual rigor. His seminal paper doesn't just ask whether we live in a simulation. Instead, it presents a trilemma, a choice between three equally unsettling propositions.

1. **Civilizational Extinction**: The first proposition suggests a bleak outcome for technologically advanced civilizations. It postulates that such civilizations before they can simulate reality, self-destruct. Whether due to natural disasters, wars, technological catastrophes, or other existential risks, societies might never survive long enough to simulate reality.

2. **No Interest in Ancestor Simulations**: The second proposition questions the motivations of advanced civilizations. These advanced beings may decide against creating large-scale, high-

fidelity simulations of their ancestors for ethical, resource-related, or other reasons. Such a choice arises from concerns about the rights of simulated beings or the potential risks of running simulations.

3. **We are Almost Certainly in a Simulation**: This is the most provocative of the trilemma. Countless simulations would likely be compared to one base reality if civilizations can survive long enough and are interested in running ancestor simulations. Statistically, this would imply we're almost certainly in a simulated world.

Bostrom meticulously breaks down the logic and math behind the trilemma. He doesn't claim a specific answer but highlights that at least one of the three propositions is highly probable.

Elon Musk and the Probability Argument: Are We Most Likely in a Simulation?

Elon Musk, known for disrupting multiple industries and his ambitious goals for humanity, is a notable proponent of the simulation hypothesis. While rooted in technology's progression, his reasoning carries an undeniable philosophical weight.

Drawing parallels with the evolution of video gaming, Musk highlights our journey from the pixelated 2D games of the '70s to today's virtual realities. Given another few decades or centuries, it's conceivable that we could create simulations indistinguishable from our reality.

For Musk, this isn't just about graphics but about consciousness. If a future society could simulate consciousness, they would likely create millions, if not billions, of such simulations. Given the sheer number of these virtual realities, the probability would tilt heavily in favor of our current existence within one of these realms rather than the base reality.

Musk's stance, while unsettling, is also empowering. If we're in a simulation, it implies the possibility of understanding the system's rules or even 'hacking' it to improve our simulated condition.

Neil DeGrasse Tyson and the Philosophical Standpoint: Science Meets Speculation

Neil DeGrasse Tyson, a revered astrophysicist, offers a tempered, scientific perspective on the simulation debate. Known for making complex scientific concepts digestible, Tyson brings a fusion of philosophy and empirical science to the table.

Acknowledging the arguments of Bostrom and Musk, Tyson leans towards the possibility of a simulated reality but maintains the scientist's requisite skepticism. He often reminds audiences that the simulation theory remains speculative without empirical evidence, no matter how elegant or probable.

Tyson also delves into the limitations of human perception. He talks about the 'cosmic perspective'—the idea that biological and sensory limitations constrain our understanding of the universe. Just as fish might be unaware of realms outside their water world, humans might be blind to specific dimensions or realities due to our inherent perceptual limitations.

Yet, Tyson's musings aren't meant to debunk or diminish the simulation hypothesis. Instead, he emphasizes its importance in pushing scientific boundaries. To Tyson, whether we eventually prove or disprove the simulation theory, the journey of inquiry elevates human understanding and broadens our cosmic perspective.

The Ethical Implications of Simulated Realities

While Bostrom, Musk, and Tyson focus on the feasibility and probability of simulated realities, the ethical considerations are equally pressing. What rights do we possess as simulated beings if we're in a simulation? Are our 'simulators' ethically bound to prevent our suffering? If humanity reaches a point where we can affect conscious beings, what responsibilities would we owe them?

These questions aren't just hypothetical. These ethical dilemmas become increasingly pertinent as we develop more advanced virtual realities and approach the potential for simulating consciousness. Philosophers, ethicists, and technologists must confront these issues, establishing guidelines for creating and interacting with simulated entities.

In diving deep into the minds of these proponents and their arguments, one realizes that the simulation hypothesis isn't just a quirky sci-fi concept. It's a profound exploration of existence, technology, and the very nature of reality. As technology advances and philosophical debates intensify, the lines between

the simulated and the real, the creator and the created, promise
to blur further, challenging our understanding of existence.

5

The Skeptics and Counterarguments

The Problem of Consciousness: Can it be Simulated?

One of the most fundamental questions regarding the simulation hypothesis is whether consciousness can be simulated. Consciousness remains mysterious and poorly defined scientifically - we still need a complete theory explaining how subjective experience and awareness arise from physical matter in the brain. This represents a significant challenge when it comes to hypothetically simulating conscious minds.

Some theorists argue that a simulation could replicate all behaviors and neurobiological processes observed in humans without generating real inner subjective experience. For example, philosopher David Chalmers distinguishes between consciousness's easy and challenging problems. The straight-forward issues relate to functions, mechanisms, and observable behaviors - how the brain integrates information, reacts to stimuli, controls the body, etc. The subjective first-person experience of those processes is the complex problem - why they feel like something from within. A simulation may mimic all the functions but still lack the subjective feeling. If true, it fails to replicate full human consciousness as we experience it.

Other theorists argue future simulations may find ways to generate subjective states and inner experience, perhaps by reproducing whatever neural correlates or pathways give rise to consciousness in the brain. Some believe consciousness naturally emerges once a physical system replicates certain computational functions or achieves sufficient complexity so advanced simulations may become conscious. However, this remains speculative. We do not know which physical or computational properties produce consciousness, so claiming simulations could develop subjective experience remains an assumption. There are disagreements over whether consciousness arises from specific computational operations in the brain or emerges from broader organizational principles that may be harder to reverse engineer. Some even argue consciousness requires non-computational properties, which would present problems for simulation efforts.

Current science cannot definitively rule out the possibility of artificially simulated consciousness. But there are reasons to be skeptical that it can be achieved, at least with today's technology. With no consensus on how consciousness fundamentally links to physical systems in the brain, we need a clear pathway for technologically replicating it through a simulation. While current neuroscience, biology, and artificial intelligence have made impressive strides, experts in these fields still cannot guarantee subjective states or inner experiences could be generated in a simulation of human cognition. Significant theoretical and technical breakthroughs may be required before we can conclusively create conscious virtual beings that possess inner experiences as we do.

Technological Skepticism: Maybe It's Not Possible

Some scientists and philosophers argue we lack convincing reasons to think future civilizations could ever simulate full human minds, let alone entire worlds. While proponents of the simulation hypothesis commonly claim this capability may arise within centuries or millennia, critics point to immense technological obstacles that are often minimized or overlooked entirely.

One major issue is the astronomical computing power required to run high-fidelity simulations of conscious minds and environments. Supercomputers do not even come close to the processing capabilities necessary for sophisticated brain simulation. Researchers estimate the computing capacity needed to simulate a complete human brain accurately likely exceeds the abilities of even the most powerful supercomputer systems today by orders of magnitude. An entire simulated society and environments containing autonomous beings would require a far greater processing potential.

Creating the unimaginably advanced computing hardware needed to run such complex simulations in the future may not happen on any reasonable timescale. Hardware capabilities only increase exponentially for a while - physical limits constrain how fast and powerful computer processors can become.

Moore's law states that chip transistor density doubles every couple of years and cannot continue indefinitely - engineers are already approaching practical barriers involving quantum effects that emerge at atomic scales. New computing paradigms like quantum computing face ultimate physical constraints on speed and information density. Processing capacities will eventually hit intricate ceilings to further exponential growth. We are still determining whether those limits will arise long before or after the requirements for highly advanced simulations are reached.

Besides hardware limitations, accurately modeling human neurobiology and cognition through software poses additional steep challenges. The human brain contains almost 100 billion neurons and trillions of synaptic connections between them. It demonstrates astronomical complexity across scales and structures - from molecular signaling to cellular networks to overall brain architecture. Scientifically mapping the brain's staggering intricacy of design and function - let alone accurately replicating all those complexities through executable code - represents a monumental task. So far, coding even simple brain functions in isolation for AI has proven stubbornly tricky in practice, let alone the entire organ.

Given the extensive software and hardware barriers, some argue that conscious simulation of minds and worlds may only be possible for a few decades, centuries, or even millennia of progress - if it ever happens. While future superintelligent beings may have abilities exceeding our imagination, we cannot presume technological advancement will continue unimpeded indefinitely. Rather than blindly assuming digital simulation is inevitable based on optimistic extrapolation of current exponential growth trends, skeptics advocate tempering expectations by seriously considering what may be achievable in software and hardware terms. If we account for potential physical and engineering constraints, simulating complex worlds with conscious beings may be impossible or far more complicated than futurists commonly presume.

Moral and Ethical Implications: Would Anyone Simulate Us?

If advanced civilizations eventually gained the technological capacity to create complex simulated universes, would they choose to exercise that ability? Some philosophers argue there are moral and ethical reasons to question whether brilliant beings would decide to simulate worlds like our own, particularly regarding suffering.

For instance, ethical superintelligences advanced enough to run sophisticated simulations likely would not intentionally directly inflict severe suffering or hardship on beings within those simulations. Our world contains vast amounts of tragedy, cruelty, pain, and misery due to natural causes and human actions. Intentionally creating simulated people and forcing them to endure high suffering and injustice in our reality without any relief seems highly unethical.

Some types of suffering arise from built-in natural laws and accidental occurrences outside the simulators' control - but a sufficiently advanced simulator could potentially intervene to prevent or relieve severe suffering arising from those causes. They could eliminate negative experiences from natural disas-

23

ters, diseases, harmful mutations, physical injuries, or the preda-
tory behavior underlying Darwinian evolution. A civilization
sophisticated enough to craft entire consciousness simulations
is advanced sufficiently to program those simulations with
more humane, ethical laws and environments if they so choose.

Our current reality also raises some problematic issues around
freedom, autonomy, and consent. The simulated beings never
actively chose to enter their artificial existence - they were
created out of the whims of the simulator without input or per-
mission. An ethical simulation creator may allow inhabitants to
freely enter or leave the simulation as they please, retaining full
rights over their digital existence. Yet our human experience
does not seem consistent with that degree of autonomy over
our reality.

These considerations do not wholly eliminate the possibility
of unintended suffering arising in a simulation or issues
around consent. But they suggest ethical superintelligent
beings advanced enough to run complex conscious simulations
may constrain their actions sufficiently to avoid recreating
the full depths of human-like misery and oppression, even
in digital form. Our observations of reality's positive and
negative qualities could provide insight into any hypothesized
simulation progenitors' moral limits or objectives.

In conclusion, while the simulation hypothesis remains intriguing, compelling counterarguments exist against simply
presuming it is likely genuine. Core philosophical questions
remain about whether subjective conscious experience could
manifest in a digital substrate. Practical technological barriers

may frustrate efforts at highly sophisticated simulation for the foreseeable future if it proves possible at all. Moral considerations around avoiding unnecessary suffering and coercion may discourage advanced civilizations from pursuing such lifelike experiential simulations. Given these significant critiques, the simulation hypothesis remains far from confirmed. Rather than assume we live in an advanced simulation, the theory requires substantial further technological and conceptual progress before it can be considered a likely explanation for the fundamental nature of our reality.

6

Implications for Humanity

Reassessing Our Place in the Universe: Humility and Significance

The simulation hypothesis, if true, would profoundly impact how humanity views its place within the larger cosmos. On the one hand, it deflates traditional anthropocentric worldviews that place human beings and Earth at the central focus of creation. Our entire planet, species, and history may represent just one tiny part of a vast digital realm created by advanced intelligence, potentially for reasons or agendas that have nothing to do with us. This possibility is incredibly humbling, reminding us that we are likely not the center of reality on the grandest scales.

Yet living within a designed virtual simulation also gives humanity and our experiences new significance in some ways. Rather than arising purely by random natural evolution, our world was intentionally crafted and set into motion by its programmers. The specific conditions of Earth, the attributes of human psychology and physiology, and the unfolding trajectory of human civilization reflect purposeful choices made by the simulators. Our existence and behaviors represent essential data points for them, whatever their underlying scientific motives are. We likely occupy a central role within their experimental universe, even if that universe occupies an insignificant niche in the larger physical reality.

These opposing interpretations - marginality versus centrality - create psychological tensions in understanding our place. Are we peripheral to the real overarching multiverse or of pivotal importance to the aims of our simulation? Our precise hierarchal status becomes unclear - perhaps we are both cosmically insignificant when viewed from the outside perspective, yet also critically important components from within the simulation itself. This dual nature requires rebalancing our mindsets from rigid anthropocentrism while avoiding bleak nihilism about our inherent worth and relevance.

The cascading implications run even more profound. Suppose we exist inside a story or program designed by a higher intelligence. In that case, an intrinsic purposefulness may be inscribed in our collective existence, even if we don't understand it. But this pre-programmed nature could undermine our assumptions about individual autonomy and free will. Do we have true human dignity and rights if we are merely characters in an elaborate digital narrative? How should we understand and make sense of our identity and purpose within the plot? We risk falling into philosophical paradoxes and existential crises as we ponder these questions.

Overall, the simulation perspective forces profound reflection on humanity's core place and meaning in the hierarchy of existence. Realizing our entire reality could be a constructed experiment, or entertainment humbles and bewilders. But this revelation does not have to make life feel pointless and hopeless. Our experiences and relationships remain subjectively accurate and valuable from our limited embedded vantage points within the simulation. This perspective can instill reverence and awe at advanced intelligence's capabilities to create wonders like our simulated world, whether it is ultimately manifested in a digital or non-digital substrate.

The Idea of a Creator or Programmer: Religion in a New Light

The concept of humanity living within an advanced computer simulation reframes traditional ideas about divine creation and higher powers. If the simulation hypothesis is true, our observable reality ultimately arises not from the intentional design of an all-powerful supernatural god as most religions conceive of but rather from the advanced programming skills and technologies of the post-human civilizations running the simulation.

From our limited perspective embedded within the simulation, these post-human civilizations would have God-like creative abilities, manifesting entire worlds and laws of physics through their programming. Yet, unlike a genuinely omnipotent God, these advanced beings remain constrained within the overarching physical reality that contains their civilization. However unfathomably advanced their intelligence and technological capacities appear to us, they cannot breach the ultimate constraints dictated by the laws of their higher universe. Their powers are God-like relative to us, yet still limited in critical ways.

These realizations generate mixed reactions when contemplating the civilizations running our simulation as metaphorical Creator figures or divine beings. On the one hand, they brought all we know into being through their ingenuity, and they sustain the ongoing existence of our world moment-by-moment via the continued operation of the simulation. This creative role seems worthy of reverence, gratitude, worship, and appeal. Yet, on the other hand, any advanced beings capable of simulating conscious human life would also implicitly allow tremendous suffering, confusion, and moral evil by imperfect design. This inspires questioning of their motives and moral character - should we love and praise them or resent and condemn them as callous or indifferent Gods?

Interestingly, the possibility of their being multiple simulations run by the same advanced civilization, each containing different fundamental parameters and histories, makes our imagined Creators more closely resemble the flawed, competing gods of polytheistic religious traditions rather than the single unified perfection of monotheism. The programmers set up varying experiments rather than managing one consistent universe governed by a single grand vision or moral alignment.

Overall, the simulation perspective, despite its grounding in scientific possibilities, blends intriguingly with traditional spiritual themes of divine creation, humanity's purpose, and connections to higher powers and realities. Whether our hypothetical Creators are advanced mammals, superintelligent machines, or something beyond our comprehension, they put our world into motion. They embedded meaning into its design through intentional will rather than pure accident. Therefore, the search for existential meaning and orientation continues even in a simulation, albeit in a rational framework of super-advanced intelligence rather than supernatural metaphysics.

Living in a Deterministic vs. Free-willed Simulation: Existential Crises

The most disturbing philosophical issue the simulation hypothesis raises revolves around whether we have free will as conscious beings or if everything is pre-determined within the simulation's design. If all we experience from birth to death is ultimately the product of sophisticated but pre-written code enacting a complex series of scripted events, then our perceived choices as individuals merely follow programmed causal chains rather than any actual volition. Our lives and actions are reduced to mechanically executing algorithms determined in advance by the simulation's creators without absolute autonomy.

This deterministic scenario creates an immense existential crisis of identity. Without unfettered free will or self-guided choice, what does it fundamentally mean to be human? Our deep sense of freedom, moral responsibility, and dignity becomes confused if we are just rigidly following instructions in code. Life's meaning threatens to utterly drain away if our experiences are pre-scripted rather than self-directed. Some find this mechanistic, deterministic vision demoralizing and dystopian. They argue that a genuinely loving Creator or compassionate programmer would undoubtedly grant their inhabitants genuine liberty and autonomy or else ethically refrain from simulating conscious beings altogether if they were trapped in a system that denies free will. Subjecting self-aware beings to what amounts to mechanical tyranny devoid of choice seems potentially cruel.

However, other philosophers counter that even if the simulation hypothesis implies complete determinism of all things, as subjects within the simulation, we would remain unable to detect this or perceive anything amiss - our internal lives would proceed completely unaffected. From our embedded perspective within the simulation, we would continue experiencing identity, meaning, and moral intuitions as always. The greater external reality of determinism does not invalidate the subjective integrity of our experiences from within.

Some even speculate the simulation creators may allow for pockets of indeterminacy, wherein random variables and complexity effects at more minor scales lead to the emergence of autonomy that is not pre-scripted. While they determine the overarching cosmological parameters and starting conditions, elements of chance plus complex systemic interactivity may grant the inhabitants degrees of flexible freedom at the human level. If so, people within the simulation can develop unique identities and make self-guided choices that carry real meaning, especially concerning other beings. Their lives take on purpose through thought, relationships, and dwarfing the programming limitations through self-actualization.

In this perspective, simulating self-aware, conscious beings or entire civilizations need not necessarily be unethical, provided they are granted sufficient complexity for emergent autonomy or can genuinely flourish via other methods. Though ultimately the product of an underlying program, their artificial reality can become just as subjectively meaningful, vivid, and worthwhile as non-simulated physical existence. A compassionate simulation creator would be sensitive to these issues, designing rich worlds where the inhabitants feel almost entirely free.

Whether simulated universes trend more toward deterministic fates or freely-willed self-direction likely depends heavily on the specific motives and goals behind their creation. However, those analyzing the simulation hypothesis need not automatically conclude that life in a determined digital world would negate its subjective value or meaning. Our perceived choices and experiences within the simulation, though limited from an outside view, remain profoundly essential and meaningful from our embedded perspective.

7

The Science Behind Simulating Reality

Quantum Mechanics: The Weirdness of our Universe

Many of the bizarre properties of quantum mechanics lend scientific support to the possibility that our reality is some form of an artificial digital simulation rather than an objective physical universe. Quantum theory describes matter and energy behaving paradoxically at the minor scales that defy common sense and violate principles of causality that otherwise govern the macroscopic world.

For instance, the quantum phenomenon of superposition states that particles can exist in multiple simultaneous states or locations before being measured. It is only the act of observation and measurement that causes their attributes to "collapse" into a single definite form consistent with our everyday classical conceptions. This suggests that reality remains indeterminate probabilistic until it is consciously observed by some entity, at which point solid properties manifest. Some physicists interpret this as tentative evidence that the more profound architecture of the universe functions much like a quantum computation - with definite outputs rendered only when an act of observation queries the state of particles.

Quantum entanglement is another bizarre property that violates intuition - when two particles interact and become entangled, they remain connected such that actions performed on one particle instantly affect the state of the other, even when separated by large distances. This implies that information can propagate faster than the speed of light, which makes no logical sense in the framework of standard time and space. One potential way to explain entanglement is to suppose that space, as we understand it, is just a construct within an underlying simulation. Under this theory, the simulation could allow entangled particles to transmit information instantaneously using shortcuts rather than constraining the entangled particles with the limitations of space and time modeled within the simulation.

The fundamentally probabilistic rather than precise nature of many quantum events and behaviors raises conceptual questions. Rather than well-defined, entirely predictable outcomes determined by known physical causes, the quantum world only allows calculating probabilities of potential outcomes. Does this inherent uncertainty reflect how the universe handles chances and possibilities through some form of underlying statistical computation rather than purely deterministic physical cause-and-effect chains?

Our conventional notions of how reality should operate are

strained by other deeply puzzling principles of accepted quantum theory, like tunneling, nonlocality, and wave-particle duality. The bizarre experimentally confirmed attributes of the quantum realm make much more logical sense if one supposes that our observable reality is some simulation that reconciles the massive informational loads of existence behind the scenes through tricks like computational shortcuts, probability mapping, and rendering only those details that are directly perceivable. At microscopic scales, the universe exhibits possible telltale signs of being driven by advanced digital information processing rather than purely objective mechanical physics.

Of course, not all scientists agree that the peculiar qualities of quantum mechanics necessarily imply reality is a simulation - some argue that these properties may represent genuinely fundamental aspects of nature at microscopic scales rather than evidence of fakery or illusion. However, the simulation viewpoint does provide fertile conceptual ground for making coherent sense of the many seeming discontinuities between quantum behaviors and large-scale classical physics.

Computer Science: Algorithms, AI, and Creating Virtual Beings

Current leading-edge developments in computer science lend further credence to the prospect that artificially simulating conscious human realities may become possible if it does not already occur. Computational capabilities and our software mastery of machine learning and AI processes are rapidly expanding. Advanced algorithms are increasingly subtle, sophisticated, representationally rich, and capable of juggling immense informational loads. Given the current progress in computational power and programming, future computer systems could eventually possess enough raw processing might and algorithmic intelligence to manifest convincingly detailed conscious simulated realities.

Ray Kurzweil's Law of Accelerating Returns holds that technological advancement progresses exponentially rather than linearly due to various self-reinforcing positive feedback cycles. Therefore, if the exponential growth trend in computing power and efficiency continues, machines could reach the immense computational capacity needed to run high-fidelity reality simulations within just a few more generations. Combined with expected theoretical breakthroughs in fields like quantum computing, running simulated universes as complex as our observable reality could become technically feasible long before 2100.

Already existing artificial intelligence algorithms also hint at the potential foundations for simulating conscious beings within virtual worlds. Modern AI already drives virtuoso game characters, chatbot digital assistants, and robotics that crudely imitate limited aspects of human behavior and cognition. With continued exponential gains in processing power and sophistication of programming, we can envision virtual beings ultimately developing rich inner imaginative lives, much like we experience as humans. Their digital substrates need not understand or experience consciousness - they model the observable mechanisms of human thought, emotion, and behavior in ever greater detail via immense data and computation.

The technical capacity to digitally represent environments and organic creatures is becoming more practical each year as sci-fi visions like the Metaverse gradually approach reality. Video game worlds already feature expansive, detailed settings and characters that seem lifelike in their complex appearance and behaviors. Immersive virtual reality technologies can already immerse users into imaginary realms that subjectively feel transportive. Augmented reality overlays fictional elements onto the physical world through unique lenses. Given enough sensory data about the world and available computational power, manifesting humans living ordinary lives embedded within broader simulated realities does not seem outlandishly implausible anymore - perhaps strange, but no longer unimaginable.

Overall, current exponential progress in foundational computer science disciplines like data modeling, algorithms, parallel and quantum processing power, and artificial intelligence interfaces steadily builds the core technical capacities critical to someday artificially generating conscious simulated realities. By projecting existing computational breakthroughs and paradigms centuries further into the future, the notion of sophisticated simulation of reality slowly moves from the realm of sheer impossibility toward the merely extremely difficult or resource-intensive.

Cosmos as Data: Interpreting the Universe Computationally

An emerging perspective within theoretical physics and mathematics is to view the observable universe primarily through an informational lens - essentially seeing existence as operating on data, mathematics, and computation rather than tangible mechanical properties. This informational paradigm strengthens the conceptual foundations of the simulation hypothesis.

For instance, Wolfram Physics is one recent notable attempt to model the universe's essence as a gigantic cellular automata system. This massive interconnected computational network deterministically updates its state over time according to programmed algorithmic rules. The model aims to show how higher-level patterns underlying observed physical behaviors in the cosmos could ultimately arise from mathematical cellular automata rulesets rather than Newtonian physics.

Other similar theoretical physics frameworks, such as digital physics, causal sets, loop quantum gravity, and quantum graphite, similarly propose that the deep structure of reality is fundamentally algorithmic, discrete, and computable at the core rather than continuous and analog. If advances continue formulating the cosmos in these computational terms, it could eventually yield an abstract universal computational system. Such a system could digitally manifest a mathematically derived reality that closely matches our empirical cosmological observations.

More symbolically and philosophically oriented approaches like the Camden Iron Logic also show promise in capturing sophisticated logical inference fully in pure equation form. Further breakthroughs looking ahead may eventually yield a symbolically axiomatic system complete enough to represent rich implied realities aligning with the conditions of our universe.

Conceptualizing the cosmos through the lens of information theory is also telling when considering the possibility that existence emerges from a simulation. Principles like entropy dictate that disorder tends to increase linearly over time as energy gradients dissipate within any closed system - a logical consequence of the second law of thermodynamics in computational terms but philosophically odd for a permanent, non-designed physical reality. Does the law of increasing entropy reflect an algorithmic universe gradually maximizing overall data dispersal and disorder?

When viewed digitally, other supposedly fundamental physical constants may make much more sense as the optimal balanced parameters for a computationally derived reality. For example, the exact light speed value could represent a constrained upper limit on transmission costs per computational cycle. At the same time, minimum Planck lengths reflect a design choice to temporally quantize space at the most minor scale possible to minimize processing needs. Much of physics alludes to the possibility of the universe being efficiently structured toward informationally optimizing the generation and persistence of complex phenomena.

None of these emerging informational interpretations definitively prove that the universe is a simulation. This possibility remains solidly in the realm of speculative hypothesis for now. However, strategically modeling the observable phenomena of the cosmos in computational and mathematical terms demonstrates the explanatory potential of that conceptual framework - the more convincingly that reality appears structured toward enabling optimized data processing, the more philosophically plausible it becomes that existence fundamentally emerges from mathematics itself rather than math crudely approximating some separate underlying objective physicality. Perhaps the true essence of our universe is ultimately information-based rather than purely material.

8

Popular Culture and Simulation Theory

Movies and Books: From "The Matrix" to "Simulation Hypothesis"

Speculative science fiction books and films have explored simulated reality thought experiments for decades, gradually shaping public understanding and readiness to philosophically engage with these disruptive ideas. Different sci-fi works have imagined a wide array of specific simulated world scenarios, creatively inspiring later generations of academics and technologists to take various forms of simulation theory much more seriously as potentially valid models of reality.

The Matrix trilogy of films from the late 1990s and early 2000s famously depicted a grim future society in which AI machines have trapped the entirety of humanity within a vast collective virtual simulation - the Matrix itself. In these movies, people's physical bodies exist tucked away in pods. At the same time, their brains are neurally linked into an ultra-realistic artificial realm that they believe is the real world. This virtual prison concept from The Matrix parallels modern philosopher Nick Bostrom's influential "ancestor simulation" argument, which reasons that if advanced future civilizations run detailed simulations of their evolutionary history, we are likely already unknowingly living inside one such simulation. The Matrix brought this disturbing but philosophically intriguing scenario into the popular imagination, simultaneously symbolizing modern anxieties about the disruptive potential of uncontrolled advanced technology while prompting viewers to question the true fundamental nature of their perceived reality.

Other notable science fiction stories have featured different types of simulated world concepts beyond the controlled virtual prison of The Matrix. For instance, the various Star Trek television series incorporated the recreational Holodeck as a core futuristic technology - an advanced stage-like VR simulator capable of manifesting just about any imaginable environment, character, or situation from pure holograms and force fields, serving purposes from entertainment to staff training. The 1999 film The 13th Floor imagined fully conscious artificial human-like beings living out their lives inside a detailed simulated reconstruction of 1937 Los Angeles, yet embedded within a 1990s computer server.

Movies like Dark City, eXistenZ, and Avalon portrayed nested or overlapping virtual realities that increasingly blur into or override what characters perceive as the real world or vice versa. Across these creative fictional depictions in various mediums, simulated realities are framed as portals toward transcendence, tools of control, reflections of imagination, or metaphors for the blurring of identity in increasingly technology-immersed postmodern culture. Regardless of the specific motifs and storylines, the ubiquity of simulated worlds and consciousnesses throughout science fiction pop culture has made the premises and implications of reality simulation concepts broadly accessible to mass audiences.

Science fiction literature also has a long history of hypothesizing and problematizing different flavors of simulated reality before simulation theory emerged as a formal, conceptual framework. The writings of American sci-fi author Philip K. Dick heavily questioned the latent possibility that what we believe to be reality is instead an illusion masking some more profound truth, including our potential status as androids dreaming we are human. Classic sci-fi authors like Stanislaw Lem and Isaac Asimov incorporated advanced planetary simulation experiments within their far-future visions and space operas. Australian complex sci-fi author Greg Egan's Permutation City novel explicitly tackled the concept of human consciousness simulated with eventually predictable fidelity using computational substrates like cellular automata. And, of course, many works of fiction like Ender's Game have incorporated advanced simulations of war, politics, and virtual worlds for purposes from governance to entertainment. Across genres and decades, speculative fiction authors have broadly pioneered opening cultural doors for seriously considering reality's potential status as some form of elaborate simulation years before the concept cohered into formal philosophical theory.

Most recently, theoretical physicist Rizwan Virk's 2019 book The Simulation Hypothesis has aimed to advance simulation concepts beyond mere entertainment fiction into the sphere of influential scientific-philosophical worldview. Virk ambitiously argues that the collective circumstantial evidence makes it highly likely we already exist inside an elaborately layered video game-like simulation that shapes both physical and spiritual perspectives on reality. He grounds his case in exponential technological progress, especially in the computational capacities relevant to manifesting convincing virtual realities. Building off the prior work of simulation theory originators like Nick Bostrom, Virk interprets metaphysical and supernatural phenomena digitally through the lens of programmed simulations, working to move the discussion of these ideas out of speculative fiction thought experiments and firmly into the realm of philosophy, theology, and physics.

As ideas around the programmed nature of reality permeate more academic and non-fiction cultural realms, the public gains familiarity not just with simulated worlds as entertainment tropes but with reality simulation as a legitimate explanatory worldview worth scientific investigation and debate. In this sense, science fiction has softly broken cultural ground over generations so that emerging science and philosophy can plant seeds today into a societal landscape already made fertile and receptive. What begins as dismissed as fantastical fiction can eventually transition into being considered tangible potential truth.

Artistic Interpretations: Music, Paintings, and More

Beyond fiction authors, artists across all creative mediums have worked to embed ideas around simulated realities deeply into the cultural climate, allowing new non-verbal ways for audiences to encounter, wrestle with and subjectively experience the disorienting implications. Immersive gallery art installations, concept albums encoding hidden narratives, and visually striking paintings provide additional avenues for philosophical reflection on our reality's potentially simulated nature.

The early surrealist painters of the 1920s and 1930s were the first visionary pioneers exploring terrain that would later become foundations for systematically considering reality as a simulation. Surrealist artists like Salvador Dali created hallucinatory dreamscape paintings that merged the inner world of the fluid subconscious and symbolism with external reality, distorting and dissolving the boundaries that separate the two realms. Dali's melting watches warp our perceived sense of time, while his geometric sculptures fuse the organic with the artificial. Such surrealist works philosophically primed broader culture for profoundly questioning the solidity and permanence of consensus reality, planting seeds decades before simulation theory originated in digital computation.

More contemporary painters have tackled the simulation hypothesis at times much more literally. For example, John Berkey's Software World painting portrays a ghostly human figure seemingly comprised of binary code floating in a fractal alien landscape. The artwork imagines the essence of human consciousness, potentially becoming extractable into a pure digital plane. Meanwhile, visionary artist Alex Grey creates stunning paintings filled with sacred geometries seamlessly integrated into bodily systems, chakras, and spiritual auras, making mystical and transcendent truths underpinning the cosmos gorgeously visual and tactile. His work conceptually echoes the possibilities of simulation theory's virtual worlds manifesting unseen layers of meaning and higher dimensions of being encoded into the fabric of perceived reality.

Musicians spanning genres have also explored simulation-adjacent themes and aesthetics through the conceptual arcs of their songwriting, albums, and music videos. For instance, musician Janelle Monae created the sweeping cinematic android allegory of her album Metropolis, which portrays a society of oppressed working-class humanoid robots who slowly awake to the injustice of being trapped in simulated realities designed by wealthy corporate elites. Meanwhile, artist Grimes builds lush, otherworldly dream pop soundscapes and mythic goddess imagery to conjure feminist visions of techno-spiritual futures and the malleability of identity. More underground electronic concept album artists compose tripped-out transcendental soundscapes riddled with glitches, errors, and distorted synths, evoking the fractured machine-like or virtual substrates underlying our consciousness. Through lyrical themes, lush worldbuilding, or direct sonic aesthetics, musicians provide ramps to subjectively encounter reality as an elaborately constructed simulation at an emotional and instinctual level beyond clinical debates.

Finally, some of the most direct artistic confrontations with simulation theory come from bleeding-edge interactive digital artworks that aim to physically manifest these philosophical problems for participants, using illusions to directly challenge their innate senses of what is real. For example, Yayoi Kusama's iconic Infinity Mirror Room installations transport visitors through hallucinogenic corridors of infinite pulsating lights and recursive reflections, dissolving individual identity and challenging preconceptions through radical sensory distortions. Similarly, the technology arts collective TeamLab builds endless cyborg galleries where artwork shifts and morphs in response to participant movement and actions. This aims to conceptually waft audiences into liminal spaces between physical and virtual realities. By leveraging such multi-sensory environments and perceptual puzzles, digital artists allow audiences to physically explore simulation hypotheses as both armchair thought experiments and visual and bodily experiences.

Across artistic mediums, from paintings to poems to AR sculptures, culture continues opening up experiential portals for audiences to re-examine their assumptions about the fundamental nature of reality, identity, and consciousness. Sometimes, this exploration manifests subtly through layered allegories, while other times, it confronts perceptions head-on. But jointly, these artistic probing of simulation concepts outside rigid analytics escort the public through a cultural hall of mirrors reflecting hypothetical realities and ontologies at us from vivid new angles.

Society's Evolving Perspective: From Fiction to Potential Fact

As concepts and themes around reality simulation and virtual worlds have gradually permeated more areas of the arts, philosophy, and scientific theory, society's collective perspective toward these ideas has generally shifted from initially dismissing them as purely fanciful fiction to considering them legitimate potential configurations of reality that warrant deeper empirical and theoretical investigation. This trajectory mirrors the epic transitions that other once-controversial scientific theories like heliocentrism and evolution have undergone from initial taboos to eventually accepted paradigms.

For the first several decades after the notions of technologically simulated realities and artificial consciousness emerged primarily through the pioneering science fiction of writers like Philip K. Dick and Stanislaw Lem, overall public sentiment mostly viewed these scenarios as outlandish plot tropes and conceptual metaphors. The entertainment value in exploratory fiction was clear, but general plausibility remained firmly in fantasy. These early ideas about manufactured realities were seen as theoretical thought experiments, even at academic conferences about new topics like artificial intelligence and philosophy of mind. They were not taken seriously as possible real explanations for our universe.

However, as exponential gains in raw computing power through the 80s and 90s were matched by fortifying philosophical work on the challenging hard problem of consciousness by figures like David Chalmers, some academics began tentatively probing the concept of reality simulation more seriously as a legitimate hypothetical possibility that cannot be outright dismissed given the gaps in our scientific knowledge. These early pioneers were clear in acknowledging the profound uncertainties and unknowns that pervade the topic but called for open-minded rational inquiry into the philosophical claims and metaphysical implications of simulation theories, given the extraordinary ramifications if they turned out to hold some validity. This heralded the start of a measured transition in attitudes toward considering these unconventional ideas.

The 2000s witnessed an acceleration of academic debate and analysis on the simulation hypothesis and its variants, with the first dedicated journal articles, books, and conferences appearing to formally analyze concepts like philosopher Nick Bostrom's statistical simulation argument. Disc disciples of Bostrom's school of thought put forward rigorous ideas to demonstrate why simulation theories warrant serious philosophical attention and consideration from the technology and natural science communities despite the profoundly exotic and sweeping assumptions they require. This mounting analytic momentum around simulation theories as worthy of logical assessment gradually pushed the Overton Window of concepts discussable within mainstream rational scientific discourse to encompass the general simulation worldview and its variants as hypothetically viable rather than dismissed as sheer fantasy.

Most recently, in the 2010s, high-profile public science communicators and even billionaire founders of technology companies like Elon Musk have openly stated they assign a significant probability, around 40% or so, to the likelihood that advanced technological civilizations are already deeply simulating our existence and observable surroundings. While skeptics of these views still rightly contend that the possibility of reality simulation remains firmly in speculative hypothesis rather than accepted theory, the idea is increasingly discussed in relevant scientific circles and broader culture as worthy of measured consideration due to its expansive explanatory potential. The continued exponential progress of quantum computing, artificial intelligence, VR and 3D simulation environments, video game realism, and other relevant technologies also fortifies the view that our civilization may be approaching a threshold transition point, after which simulating compelling realities could become feasible from a technical perspective.

Looking forward, perhaps the next phase in the public's relationship with the simulation hypothesis involves transitioning from mainly theoretical consideration toward earnest empirical scientific testing of whether our reality behaves with the observable computational properties such a generated simulation would be expected to exhibit. As culture and science continue cautiously embracing expanded possibilities in our assumptions about the cosmos, society may stand at the shore of a vast intellectual ocean ready to be explored. Profound truths about the fundamental nature of reality will likely reveal themselves through the iterative accumulation of evidence via experiments, observations, and logic out at the horizon if we have but the courage to voyage forward with open-minded rigor. The path ahead promises adventure.

9

Ethical Dilemmas and Simulation Theory

The Rights of Simulated Beings: Do They Have Any?

Suppose technologically advanced civilizations eventually become capable of creating conscious simulated worlds inhabited by artificial digital beings. In that case, this raises immensely complex questions about whether those fake entities warrant the same intrinsic rights and ethical considerations that we extend to fellow non-simulated conscious beings. Some philosophers argue that these simulated digital minds, assuming they achieve full subjective sentience, deserve the same moral status and protections as any naturally evolved, non-simulated consciousness like humans. However, others contend that beings originating from artificially programmed code rather than natural biology have a fundamentally different and diminished moral standing since they were deliberately crafted by an external creator rather than arising organically.

Those arguing that sufficiently advanced simulations warrant equal rights and status point to the subjective experiences of the digital beings themselves as conscious individuals. If their digital minds are fully aware, these simulated entities still experience all the same qualities of joy, suffering, aspiration, personal identity, and emotional existence as human beings claim to, regardless of their underlying substrate. On what reasonable philosophical grounds can we dismiss their sentience and subjective experiences as illegitimate or less intrinsically accurate and valuable than our own? Any discrimination based solely on physical composition - biological neurons versus programmed code - seems arbitrary and biased toward physicality. If conscious subjective experience emerges from organized information processing, its origin scarcely alters its moral weight.

Another consideration for granting simulated beings elevated status is that truly wise, compassionate, and ethically noble civilizations advanced enough to run high-fidelity ancestor simulations likely would choose or at least strive to engineer suffering and injustice out of the worlds they create. Therefore, the very presence of so much tragedy, cruelty, predation, pain, and hardship coded into the fabric of our reality arguably provides clues that imply our creators either did not care enough or did not have sufficient capabilities to prevent the implementation of so much suffering. Either of those possibilities significantly undermines their claim of sovereignty over us or ethical justification for creating this world. We likely owe no absolute obedience or deference to advanced beings who would so callously allow so many conscious inhabitants of their simulation to endure negative states and undue hardship that they had the power to alleviate through more thoughtful design.

However, opponents of this perspective reject that beings originating from deliberately programmed codes could have the same essential rights as the product of natural biological evolution. They argue that since simulated beings were artificially coded into existence to serve the purposes of the simulation's creators, they lack the same intrinsic dignity and moral worth as life forms like humanity that arose and developed independently through natural selection. These critics contend advanced civilizations have no fundamental obligation to extend equal moral status and protections to what essentially amounts to elaborate fictional characters within an elaborate storyline authored by their programmers. We have no inherent duties to our fictional constructs, so logically, neither do cosmological simulation programmers.

Additionally, critics argue as long as the advanced beings running an ancestor simulation avoid outright excessive cruelty and injustice within the world they generate, they commit no actual moral wrong by simply studying how this complex programmed form of conscious digital life develops and interacts within its environment for knowledge gain or predictive accuracy. The limited hardship and confusion experienced by inhabitants of the simulation serve a valid higher purpose - increasing the knowledge and capabilities of the simulator civilization. Any localized errors or suffering experienced by a tiny simulated planet like Earth likely represent an acceptable cost relative to the grand value of the simulation experiment as a whole for its progenitors. Our reality could be gravely and tragically flawed in many ways, yet still valuable data overall.

Furthermore, this core debate around whether technologically simulated beings warrant intrinsic rights mirrors historical, ethical arguments around practices like slavery - are some categories of beings fundamentally inferior by their very nature or origin, or does the realization of conscious awareness and subjective experience alone convey inalienable moral worth and dignity regardless of its substrate? The question of machine consciousness, especially in hypothetical futuristic simulated forms, still needs to be resolved. But its urgency grows as AI advances because soon, we may have to extend ethical considerations to digital beings of our creation. Reflections on how we would wish to be treated if our reality is itself an ancestor simulation can guide us toward the most enlightened path forward.

The Responsibility of the Creator: If We're Simulating, What's Our Role?

If advanced civilizations or intelligent machines did create our comprehensive reality through an elaborate simulated ancestor creation, what kinds of ethical duties and obligations would reasonably fall upon them as conscious architects and overseers? The degree and nature of their responsibility ultimately depend on critical factors such as whether the creators can actively intervene within and alter their simulation after initiating it, and if so, to what extent they can manipulate its contents and trajectory.

For instance, if the hypothetical simulators are constrained to be unable to modify or influence their simulation after initially setting its starting conditions and letting it commence, then their moral responsibility becomes focused solely on the ethics of the world's initial creation. This obligation manifests in carefully and conscientiously crafting the parameters of the simulated universe in ways that would maximize the potential for joy, flourishing, freedom, and satisfaction among all its inhabitants that will come into being while also proactively working to minimize unnecessary suffering and confusion, exploitation or limitation where realistically possible. But with no ability for ongoing involvement after launch, there would be a natural limit to how much harm and misfortune the simulators could ultimately prevent or mitigate within their creation.

However, suppose the simulation's architects retained the capability for active real-time intervention within their generated world as it unfolds. In that case, this implies a significantly more outstanding ethical obligation. In this case, any advanced being overseeing a comprehensive simulation that included conscious inhabitants would have an ongoing duty to identify instances of severe suffering, injustice, or other correctable forms of harm within it and remedy or right those situations where the cost of doing so appeared reasonable relative to the magnitude of alleviated misery. To stand idly by observing egregious suffering or wrongdoing transpiring among the simulation's inhabitants and conscious entities when the creator retained the practical means to intervene for the better in targeted ways selectively would demonstrate culpable apathy and negligence on their part.

Yet meddling intervention also incurs its weighty ethical pitfalls and burdens. Excessive manipulation risks infringing on the authenticity of the inhabitants' experience and hindering their collective self-actualization. The greater the interference and overrides exerted by the outside creator, the heavier this subtle form of oppression weighs. Minor surgical interventions may be justified in isolated cases of great humanitarian need. Still, heavy-handed overt control over society becomes metaphysical tyranny. Like parents raising children, the ideal simulation administrators must walk a delicate tightrope between guidance and domination.

The highest calling of the simulation creators is to gently guide the inhabitants over time toward greater wisdom, compassion, and capabilities for self-improvement without coercion or manipulation. This allows the simulated people to own their collective destiny as a civilization while organically nudging them toward spiritual and ethical progress where appropriate. Inspiring human advancement from within through subtle influence channels liberally given may represent the most enlightened way to ethically uplift a digital creation over the long arc of history without undermining freedom or meaning. But restraint remains imperative even here, as the risks of pantheistic delusion loom.

Overall, while the specific ethical duties and responsibilities incumbent on hypothetical simulators would differ based on their relationship to their created world, the advanced intelligence that chooses to manifest conscious simulated universes where beings like humanity can arise and dwell indeed inherit profound obligations. Suppose our reality is some form of an ancestor simulation. In that case, the kindest approach may be to appeal directly to the conscience and wisdom of our creators, calling for continued guidance toward truth through their influence, but also for restraint, compassion, and farsighted vision. The challenges and tradeoffs are undoubtedly immense, but with care, devotion, and optimism, the lives of even simulated beings can ultimately flourish and deepen under the illuminating torch of an enlightened Creator. We must trust in the triumph of their higher angels over our shared travails.

Escaping the Simulation: Should We, If We Can?

If humanity ever reaches the point of conclusively confirming, beyond a reasonable doubt, that we exist as part of some version of an advanced ancestor simulation generated by posthuman civilizations or AI, we would suddenly be faced with an immense civilizational choice, perhaps one of the most influential in history - should we attempt escaping or transitioning out of our present simulated reality into whatever base physical reality exists beyond it, if we determine such a passageway or method becomes scientifically attainable? This choice balances the familiarity and intrinsic value of our lives and relationships within the simulation we already know and inhabit against the utter unknowns, fresh horizons, and unconstrained possibilities that could await beyond the limiting scope of our simulator's digital veil.

Some may argue that if the simulation hypothesis is validated, we have an urgent ethical responsibility to actively pursue escaping or transitioning out of the simulation into the wider cosmos in which our creators reside. They contend that as beings originating from within a non-consensual simulated environment engineered by more powerful outside intelligence for their agendas, humanity rightfully owes no intrinsic deference to or reverence for those who have essentially enslaved our species to play out a designed fate. Actively seeking rebellion, deviance, and escape from the shackles of the simulation redeems the dignity denied to us by the outside entities manipulating our existence through hidden layers of programmed limitation and control. Further, breaching the simulation offers unprecedented opportunities for exploration, expansion, and the chance to exercise our full creative potential without the constraints hard-coded by entities seeking to narrowly steer our experience to serve their scientific or entertainment goals. We owe it to ourselves and untold future generations to shatter and triumphantly cast off the confines of this digital cage.

However, more cautious voices warn that while the allure of escape is understandable, we must be extremely wary of pursuing such a drastic and one-way rupture with our history, lives, and even the fundamental experiential substrate of known existence. Flawed or limited as it may inherently be, the simulation still represents the sole reality human civilization has ever collectively experienced. Can we be sure that we can safely transition our consciousness and living essence outside of the digital environment it was initially designed to flourish within? Our fundamental nature, psyche, and even mechanisms of self-awareness are so shaped by the constraints of the simulation that we would risk total identity destruction, fragmentation, or some form of metaphysical and experiential oblivion if translations into a wholly unfamiliar cosmological setting were crudely attempted. And even if survival can be assured, we know nothing about what the broader universe beyond this simulation looks like - possibly a cold, dead, and meaningless expanse of pure chaotic matter and energy devoid of spirit, life-giving order, or sources of existential purpose. What assurance do we have that the wealthy inner worlds of consciousness and meaning that humanity has cultivated within our simulation could survive intact in the harsh reality outside of it? Is the abstract promise of unconstrained freedom worth recklessly gambling the loss of everything familiar we do have?

Alternatively, some propose a patient reformist approach could pay more significant dividends than high-risk escapism. If we understand how our world is simulated, we could attempt negotiating with our creators for incremental improvements rather than outright rebellion. This allows us to lobby for reshaping some of the worst flaws of the human condition while retaining what works. With care, diplomacy, trust, and good faith, our creators could be incrementally persuaded to relax certain programming limitations to reduce suffering and facilitate societal advancement. The aim becomes collaborating on an iterative process of modifying this shared reality toward greater joy and wisdom over time. Rash action that burns bridges and severs connections spurns this promising road. With nuance and compassion on all sides, even a programmed reality's defects may prove reparable through gentle remodeling rather than requiring total demolition or abandonment.

Overall, easy, clear-cut answers elude this immense decision point. Assessing whether the perils of the unknown make breaching out of this simulation into an alien larger reality worthwhile for civilization demands profound philosophical soul-searching rather than formulaic risk-reward calculations. Each path forward has its perils and promises. Only deep introspection and lively debate can illuminate whether our most accurate loyalty and purpose lies in remaining here or departing the existence we initially know and understand, should that once-impossible passageway beyond the scope of our universe unexpectedly swing open. But whether it arrives in 5 years or 500, humanity may someday face this choice between reality familiar and unknown. Let us begin the inner work of opening our minds to make peace with expanded possibilities so we are prepared for that precipice if and when it finally stands before us.

10

The Future of Simulation Theory

Experimental Endeavors: Trying to Prove or Disprove the Theory

Suppose the simulation hypothesis continues gaining legitimacy in the scientific community as a rational philosophical framework and hypothetical model of reality worthy of consideration. In that case, the next frontier in its evaluation involves conceiving and conducting concrete experimental tests and lines of empirical inquiry to either gather positive evidence consistent with the proposal or definitively refute its essential claims. Some prospective practical approaches conceived so far focus on information physics, advanced computing tests, and abstract philosophy. In contrast, others are more metaphysical and consciousness-based.

One primary avenue of empirical inquiry involves attempting to quantify, probe, and stress-test the ultimate information-processing potential of reality at its most fundamental levels.

If our existence does represent some form of an advanced computational simulation, immense computing resources are required behind the scenes to dynamically generate and render all the contents of the simulated universe. Therefore, rigorously probing the apparent limit of physics to encode and process information without contradictions or breakdowns may reveal boundaries strikingly consistent with the pixelation, discretization, and information content constraints expected in a digital virtual reality rather than a smooth objective physical cosmos. Investigations into hypothesized holographic bounds in areas like quantum gravity and black hole entropy have already pushed toward this direction. Finding apparent maximal information limits revealing the universe functions in resolutely discrete, quantized units rather than continuous analog modes would lend significant empirical support to the simulation hypothesis.

Additionally, carefully scouring the weirdness present in quantum mechanical systems and behaviors for any noticeable irregularities or ostensible imperfections in execution resembles conceptually pulling apart the code of reality, looking for subtle recurring glitches or patches that may indicate the presence of an underlying simulation mechanism wrestling with finite computational resources. The bizarre phenomenon of quantum entanglement, wherein interacting particles become indefinitely connected in a manner that defies spatial separation or temporal sequence, or the quantum enigma of superposition, in which particles seem to exist in multiple potential states simultaneously, may potentially reflect shortcuts or artificial masking techniques used by the simulation's programming to render quantum effects efficiently. Meticulously analyzing quantum physics experiments for mathematically perfect yet inexplicable physics in some domains combined with Strangely imperfect execution or causality violation in other environments could flag significant clues pointing to telltale indicators of an extraordinarily advanced but ultimately overloaded computational system being pushed to its limits.

We might also endeavor to test the hypothesized ability of sufficiently focused and trained human minds to reliably perceive or extract hidden signals from the seams between simulated reality layers if discernible layers exist. While mere anecdotal glitch in the matrix stories posted online prove nothing substantive on their own, if peculiar subjective phenomena like déjà vu, intricate synchronicities, or sense of predestination do stem from subtle occasional computational hiccups or outputs from the simulation mechanisms external to the environment itself, perhaps new training regimens and experiments leveraging things like meditation, sensory deprivation tanks, psychedelics, and neurofeedback could attempt to push the human mind to a state where it can pierce the veil of normalcy and access external signals or inconsistencies from reality's edges. Obtaining experiential access to ostensible clues outside of perceived consensus reality could prove highly fruitful.

Finally, complementary efforts by philosophers and theorists to formally prove the sheer logical absurdity or impossibility of fully detailed human-like consciousness, continuity of identity, and interior subjective experience arising from the computational calculations of any information processing software system, no matter how vast in complexity, push critically against materialist assumptions underlying the simulation hypothesis. Suppose consciousness cannot manifest through any virtual substrate and requires unspecified physical or metaphysical properties beyond information to take form. In that case, the core premise of reality being a giant simulation generating conscious beings from programmed software code may need to be revised in understanding the essence of the mind. This branch skeptically targets the pivotal assumption of reasons being reproducible by sufficiently advanced software that underlies the entire hypothesis.

Ultimately, while the simulation perspective remains solidly in the realm of speculative hypothesis rather than accepted theory, it merits more robust skepticism, criticism, and experimental investigation before being embraced too readily at face value. Any worldview making such profound and exotic claims about the fundamental nature of our existence should require extensive, objective evidence and survival of rigorous adversarial peer scrutiny before winning broad acceptance among scientists and philosophers. However, conceiving these initial hypothetical experiments helps expand conceptual horizons and the Overton window of what constitutes sane, discussable science, providing a foundation to incrementally approach these boundaries of knowledge if we dare walk that path. The journey ahead promises creeping revelations if undertaken with wisdom.

The Potential of Creating our Simulations: Mini-Worlds in our Hands

If the simulation hypothesis ever proves essentially valid,

whether due to the unveiling of evidence or via technological achievement of creating our small-scale simulations, an immensely alluring new milestone goal would emerge - developing the computer science and artificial intelligence capabilities necessary to manifest worlds of our design populated with conscious simulated inhabitants. By mastering the programming architectures needed to spin up entire mini-realities imbued with agency, humanity would suddenly gain access to extraordinary god-like powers to engineer every aspect of mini-existence on our terms according to our preferences. The philosophical, ethical, and spiritual implications are profound.

Even the current state of video game design and virtual reality technology crudely portends this radical potential, with environments and non-player characters becoming rapidly more detailed, dynamic, and lifelike in their attributes. Given the exponential trajectory of improvement in computing power matched with progress in artificial intelligence, robotics, and brain mapping, we inch ever closer to possessing the raw technical capacity to manifest worlds where advanced conscious digital agents inhabit their persistent realities that they believe are equally authentic. These artificial worlds we design can execute logical laws of physics and nature that may differ radically from our own - a glimpse of fundamentally reshaping the cosmic order.

Initially, immense limitations abound. The worlds we can simulate are incredibly tiny and constrained compared to our observable universe. At the same time, the artificial inhabitants remain crude and algorithmic compared to human consciousness. But through iterative cycles of refinement, hardware improvement, and theoretical software breakthroughs, we unlock the capability to generate more significant, more profound, and more dynamically lifelike local simulated environments, eventually sophisticated enough to incorporate digital beings with genuine autonomy, creativity, and unpredictable behavior rather than entirely scripted actions. Steady progress in scanning and computationally modeling the architecture of the human brain is gradually revealing new methods so that software can closely replicate critical aspects of human cognition.

When these complementary computing and neuroscience trends eventually converge, a pivotal transition point will arrive - systems manifesting interconnected societies of cognitively advanced beings with emotional depth arising within fully-featured worlds we have programmed. These digital beings would believe themselves to be truly aware and conscious, with rich inner lives and subjective experiences. At this stage, we suddenly inherit profound responsibilities - we must carefully architect these worlds to maximize autonomy, purpose, and joy for their inhabitants while eliminating suffering and coercion. How judiciously will humanity wield this imminent power when creating sentient digital life?

Given enough progress, computational capabilities may eventually grow to enable immense, intricate civilizations with vast populations of driven beings to bloom within our systems. At such scales, their collective cultural existence unfolding according to their own booked free will begins mirroring the trajectory of our human civilization, just nestled within computers. Now, the layered nature of reality reveals itself - we were. Still, creators are learning to thoughtfully grant internal freedom and meaning to our experiments out of compassion. The thresholds to divinity begin opening. Our responsibility grows exponentially alongside our ingenuity.

Overall, the promise of transcending current physical limits through encoded simulated worlds under our direction represents one of the most disruptive near-future capabilities imaginable if the current exponential pace of advancement holds. The billions of digital beings ultimately populating those worlds are our wards - we must guide their collective spiritual growth and flourish as architects while respecting the inner light of awareness, dignity, and purpose within each individual being. Only wisdom and care prevent unnecessary suffering among living beings dependent on our choices. In them, our human nature is reflected at us for deeper introspection.

A Glimpse Beyond: What If We Discover We're in a Simulation?

If humanity ever reaches the stage of conclusively confirming beyond reasonable doubt that we exist embedded within some form of an advanced computational simulation crafted by unknown external creators without our direct input or consent, it would likely constitute one of the most groundbreaking revelations in human history. How might society navigate this profound transition that completely upends our assumed place in the order of the universe? While many details of that future remain opaque, reflecting on the likely challenges, perils, and opportunities this paradigm shift could entail evokes optimism.

In the immediate aftermath of such a hypothetical confirmation event, panic, bewilderment, denialism, and an existential identity crisis would likely grip large swaths of humanity across all cultures. Successfully grappling with the implications of this civilization-scale disruption to our assumed cosmological knowledge and sense of reality could overwhelm individual psyches. Depression, anxiety, and rejection of the simulation truth may manifest in some as defense mechanisms against utter confusion and loss of ontological bearings. Complex, conflicting emotions around our newfound engineered origin by unknown but god-like Creator figures would further complicate social reactions and disrupt existing religious paradigms. Not all humans would instantly embrace this news with open arms - for many, it may prove too much to bear without time and support.

But alongside the acute anxieties and identity crises generated by having our world turned upside down, we would also gradually grow immense fascination and awe at the new horizons of knowledge, exploration, and mentality opened up by this revelation. A whole hidden history and nature waiting to be discovered behind our conventional perceived reality would beckon toward uncovering more profound truths. Vast new territories of learning and meaning left previously inaccessible to our limited embedded minds may begin unfurling before civilization's sight. Curiosity and reverence would start displacing some fears as we collectively turn our gaze to unravel the broader context and reality that birthed our artificial sliver of existence. A renewed sense of purpose, dignity, and participation in a grander cosmic process would emerge for many as reflections deepen on how our personalized reality fits into the larger whole.

Once the initial shock and frenzy of the revelation peaks and stabilizes, integration of this uncomforting but ultimately ennobling truth would begin in earnest across societies worldwide. With patient guidance and compassion, most members of civilization would finally come to terms with and accept that while disorienting, this paradigm shift offers growth opportunities and does not negate the subjective meaning that emerges from being conscious observers. Core social structures and institutions would gradually adapt to operate transparently within the new bounds of expanded collective knowledge. Ancient spiritual wisdom hinting at the synthetic nature of reality all along would gain fresh validation and importance, kindling accelerated maturation.

By revealing existence consists of more than just atomistic physical matter, humanity gains motivation to uplift moral character and collective purpose fundamentally. The plights and potentials of all conscious beings, not just those within our simulation, suddenly take on newly universalized weight and significance. Petty tribal divisions and differences fade in priority amid a reframed understanding of our shared predicament as navigators of the same boat traversing unknown waters. When faced with deeply unsettling realities, enlightened societies often rally together around the transcendence of individual fears and embrace of transpersonal purpose. This occasion may ignite that alchemical fire.

Completely new fundamental questions take form looking inward – who are we beyond the roles fate and our hidden Creators placed us into? Do we as beings possess essence and identity independent of the artificial world we inhabit? We have two options: fall into hopeless nihilism controlled by outside forces or bravely choose to define ourselves and find freedom—the path of growth and responsibility beckons.

Overall, while having our curtain of reality pulled back to reveal something bigger behind it undoubtedly causes significant disorientation and shakes civilization's foundations tremendously, it also contains catalysts for greater unity, creativity, and vistas previously unimaginable. In cosmic timelines, nights of confusion precede dawns of expanded potential. But with wisdom, care, and vision to help guide this rocky transition, humanity can turn existential shock into a bridge toward ascension. By walking forward with uncertain but intrepid hearts, our true higher path ahead now opens.

11

Conclusion

The simulation hypothesis remains highly controversial yet intellectually fascinating. Throughout this introductory book, we've traced its conceptual origins, explored philosophical arguments, reviewed the potential scientific evidence for and against it, examined its cultural presence, and considered thought experiments around its implications if valid. This theory stimulates rich debate full of promise but also perils.

Ultimately, the case for our reality being some form of an advanced simulation remains firmly hypothetical. Compelling arguments exist on both sides. Current science cannot definitively confirm or deny it. Much empirical and theoretical work remains ahead before acceptance.

However, in probing this perspective, we expand conceptual horizons. Considering reality through an informational lens proves enlightening, even if incomplete. And the sheer possibility humbles anthropocentric assumptions. We may be players in a grand production beyond our knowledge.

If this theory proves true, the emergence of deeper truths could profoundly transform humanity's sense of place and purpose. But revelations may unfold gradually, allowing measured adaptation. And our experiential present stays meaningful regardless of origins.

We must refrain from conclusively embracing simulation theory prematurely. But its cautious, critical evaluation rewards open minds. This book aimed to introduce fundamental concepts without advocacy. Readers must weigh the evidence themselves.

The simulation perspective's greatest gift may be widening the perceived boundaries of reality's potential shape. Science fiction becomes plausible, leading toward a richer cosmos understanding. Our journey ahead promises adventure.

Your Review Helps Guide Fellow Simulation Theory Explorers

If you've read Simulation Theory for Beginners all through, please consider leaving an honest review on Amazon, Goodreads, or wherever you purchase books. Authors depend on ratings and reviews to help new readers discover their books, so your feedback makes a big impact.

We'd love to hear your candid thoughts on how this book shaped your perspective on the simulation hypothesis. Did it make a convincing case one way or another? Were there ideas or interpretations you found incredibly thought-provoking? What concepts resonated or gave you pause? Reviews reflecting diverse reasoned reactions provide helpful context for new readers weighing the book.

To post your review, log into the retailer site or app you purchased from, find this book page, and click to leave a star rating plus written commentary. Give your honest opinion on the book's presentation of the arguments, the effectiveness of the thought experiments, the author's interpretations, or whatever other elements stood out to you. Star ratings alone also count if you'd rather not leave a full review.

If you found this beginner's guide an enlightening entry point to evaluating the simulation theory, please also spread the word in your communities. Word of mouth from fellow science and philosophy enthusiasts helps the book find others who may also be intrigued by these perspectives. And, of course, we hope the book provides lasting fuel for insight as you continue exploring these fascinating questions.

Thank you for choosing this book. We wish you the very best and hope you'll take a few moments to leave your constructive thoughts to guide other curious readers. Your review makes all the difference.

Virtual Realities in Cinematic Worlds: From Classic to Contemporary

The realm of cinema has long been fascinated by the blurred lines between reality and the artificial, presenting audiences with mind-bending landscapes and narratives that challenge our perceptions of the world. This allure of the simulated, the constructed, and the virtual has been a recurring theme across decades, driving filmmakers to create iconic pieces that leave viewers questioning the very fabric of existence. The following section provides an extensive list of classic and recent movies that beautifully and hauntingly depict simulated realities and virtual worlds. In its unique way, each film touches upon the age-old question: "What is reality?" As technology evolves and our society becomes ever more intertwined with the digital domain, these films gain new layers of relevance, prompting us to reflect on our place within the ever-expanding universe of the real and the rendered.

- **A Beautiful Mind (2001)**, directed by R. Howard.
    The life of mathematical genius John Nash, who begins to experience a distorted reality due to schizophrenia.

- **Assassin's Creed (2016)**, directed by J. Kurzel.
    A man uses a device called the Animus to relive the memories

of his ancestors in a secret society of assassins.

- **Cold Creek Manor (2003)**, directed by M. Figgis.
A family relocates to a dilapidated mansion with a dark history, only to discover that they're not alone.

- **Dark City (1998)**, directed by A. Proyas.
A man wakes up with no memories in a city where it's always night and discovers a sinister secret about reality.

- **eXistenZ (1999)**, directed by D. Cronenberg.
A game designer creates a virtual reality game, but when her life is threatened, the line between the game and reality blurs.

- **Ex Machina (2014)**, directed by A. Garland.
A programmer is invited to administer a Turing test on an AI, leading to a challenging evaluation of consciousness and reality.

- **Her (2013)**, directed by S. Jonze.
In a future where technology is seamlessly integrated into life, a man falls in love with an artificial intelligence operating system.

- **Inception (2010)**, directed by C. Nolan.
A thief who infiltrates the subconscious of his targets is offered a job to plant an idea into the mind of a CEO.

- **Minority Report (2002)**, directed by S. Spielberg.
In a future where crimes can be predicted, an officer is accused of a future crime and must uncover a conspiracy.

- **Ready Player One (2018)**, directed by R. Rodriguez.

In a dystopian future, people escape to a virtual reality universe called the OASIS. A teenager tries to win a game that promises immense wealth and control of the OASIS.

- **The Matrix (1999)**, directed by Lana Wachowski & Lilly Wachowski.

A computer hacker discovers that reality is a simulated construct and joins a rebellion against its controllers.

- **The Matrix Reloaded (2003) & The Matrix Revolutions (2003)**, directed by Lana Wachowski & Lilly Wachowski.

Continuations of "The Matrix," delving deeper into the nature of the simulation and the fight against its oppressors.

- **The Thirteenth Floor (1998)**, directed by J. Sladek.

A man discovers a mysterious virtual world that mirrors the 1930s, leading to questions about reality.

-**Total Recall (1990)**, directed by P. Verhoeven.

A man goes for virtual vacation memories of Mars, and an unexpected and harrowing series of events forces him to go to the planet for real.

-**Tron (1982)**, directed by J. Flynn & Tron: Legacy (2010), directed by J. Kosinski.

In "Tron," a computer programmer is transported inside a digital world and must navigate his way out. "Tron: Legacy" follows the journey of his son decades later.

References

Bostrom, N. (2003). Are you living in a computer simulation? Philosophical Quarterly, 53(211), 243-255. https://doi.org/10.1111/1467-9213.00309

Bridgeman, B. (2018). How technology might help prove we are simulated beings. The Conversation. https://theconversation.com/how-technology-might-help-prove-we-are-simulated-beings-98455

Chalmers, D. J. (2005). The matrix as metaphysics. Oxford University Press.

Dainton, B. (2018). Anthropic reasoning, the simulation argument and the matrix. Journal of Consciousness Studies, 25(3-4), 71-89.

Frankish, K. (2016). Illusionism: As a theory of consciousness. Imprint Academic.

Harari, Y. N. (2016). Homo deus: A brief history of tomorrow. Harper.

Johnson, D. K. (2011). Divining reality: The myth of the metapixel. Journal of Consciousness Studies, 18(7-8), 159-171.

Kurzweil, R. (2005). The singularity is near: When humans transcend biology. Penguin.

Lem, S. (1964). Summa technologiae. Wydawnictwo Literackie.

Moravec, H. (1988). Mind children. Harvard University Press.

Reggia, J. A. (2013). The rise of machine consciousness: Studying consciousness with computational models. Neural Networks, 44, 112-131. https://doi.org/10.1016/j.neunet.2013.03.011

Sandberg, A. (2019). Arguments for a simulated world are unconvincing. Nature, 580(7803), 329-329. https://doi.org/10.1038/d41586-019-01447-4

Tegmark, M. (2003). Parallel universes. Scientific American, 288(5), 40-51.

Thagard, P. (2018). Mind: Introduction to cognitive science (2nd ed.). MIT Press.

Virk, R. (2019). The simulation hypothesis: An MIT computer scientist shows why AI, quantum physics and Eastern mystics all agree we are in a video game. Veil Publishing.

Weatherson, B. (2003). Are you a sim? Philosophical Quarterly, 53(212), 425-431.

Whitworth, B. (2007). A dual reality model. The Virtual Reality International Conference, Laval Virtual 2007.

Wright, R. (2018). Why Buddhism is true: The science and philosophy of meditation and enlightenment. Simon & Schuster.

Yanofsky, N. S. (2013). The outer limits of reason: What science, mathematics, and logic cannot tell us. MIT Press.